A
GEORGE WASHINGTON CARVER
HANDBOOK

A GEORGE WASHINGTON CARVER HANDBOOK

Dr. B. D. Mayberry

EDITOR

NewSouth Books
Montgomery | Louisville

NewSouth Books
P.O. Box 1588
Montgomery, AL 36102

ISBN-13: 978-1-60306-016-5
ISBN-10: 1-60306-016-2

Copyright © 2007 by B.D. Mayberry.
All rights reserved under International and Pan-American
Copyright Conventions. Originally published in 2003
with the ISBN 1-58838-164-1.

Second Edition, 2007

Printed in the United States of America

Contents

Preface 7

    I.   Brief Biography 11

Photos I: Early Life 15

    II.   Classified Chronology of Carver's Life and Legacy 17

Photos II: Teaching 23

    III.   USDA complex Named in Carver's Honor 25

    IV.   Carver Namesakes by State 28

Photos III: Research and Experiments 37

    V.   Carver's Achievements 41

    VI.   Comprehensive Chronology 57

Photos IV: Carver the Man 67

    VIII.   Carver Day 69

Photos V: Tributes and Honors 71

    IX.   Appendices 76

    X.   Carver Bibliography 82

Preface

Out of the darkness of slavery emerged George Washington Carver. He was born in a period when the lights of hope and aspiration for and by the Negro—coming out of the end of the Civil War and the Emancipation Proclamation—were only dimly appearing on the horizon. Yet from the meager beginning, in the shadows of slavery, he demonstrated a life worth emulating, not only for the Negro, but also for all mankind, around the globe.

Recognitions and tributes to the life and legacy of George Washington Carver have been compiled or catalogued at local, state, national, and international levels. Brief references to most of these are made in this handbook. It is hoped that this will help in the expansion of the ever-increasing knowledge of the lengthening shadows of the life and legacy of George Washington Carver, and insure that they will continue immortal in perpetuity.

This handbook is essentially a fact book on the life and legacy of George Washington Carver. More than 100 books and articles have been written and published about Dr. Carver. Most of these are listed in the bibliography of this handbook.

Individuals and agencies at local, state, national, and international levels have paused repeatedly to pay tribute to George Washington Carver. Many of these events are catalogued in this handbook.

In the Classified Chronology section, it is of special interest to note tributes paid to Carver by (1) Institutions of Higher Education, (2) State

Governments, (3) the Federal Government, especially the National Park Service and the United states Congress, and (4) public schools across the nation.

Finally, a highlight of this handbook is a list of most of the products made from soil extracts, the peanut, the sweet potato, and other plant materials.

A George Washington Carver Handbook

**Carver Bust at the
Alabama Department of Archives and History**

I.

BRIEF BIOGRAPHY

IT IS ONLY LOGICAL to expect that any discourse dealing with the life and works of George Washington Carver would begin with, at least, an overview of his early years of growth and development. Because this is designed as a handbook, no extended bibliographical narrative is included. I am recommending at least one of the more appropriate of hundreds of writings about Dr. Carver's early experiences for your review, including the following:

1. An account of Carver's early years written by Booker T. Washington in his 1911 book, *My Larger Education*;
2. "The Slave Who Freed the South," chapter one in Lawrence Elliott's book, *George Washington Carver: The Man Who Overcame*, published in 1966;
3. "Wandering in Search of Destiny," chapter one in Linda O. McMurry's book, *George Washington Carver: Scientist and Symbol*, published in 1981; and
4. *Unshakeable Faith: A Biography of Booker T. Washington and George Washington Carver* by John Perry in 1999.

These references will be adequate to show how Carver was born, grew up, and was essentially self-educated in the shadows of slavery. He emerged to become internationally recognized as a scientist and as a humanitarian. In addition, however, George Washington Carver's autobiography as written by him before the turn of 19th century is included below.

GEORGE WASHINGTON CARVER'S OWN BRIEF HISTORY OF HIS LIFE

1897 or Thereabouts

As nearly as I can trace my history, I was about two weeks old when the war closed. My parents were both slaves. Father was killed shortly after my birth while hauling wood to town on an ox wagon.

I had three sisters and one brother. Two sisters and my brother I know to be dead only as history tells me, yet I do not doubt it as they are buried in the family burying ground.

My sister, mother and myself were kuckluckled, and sold in Arkansas, and there are now so many conflicting reports concerning them, I dare not say if they are dead or alive. Mr. Carver, the gentleman who owned my mother, sent a man for us, but only I was brought back, nearly dead with whooping cough, with the report that mother and sister was dead, although some say they saw them afterwards going north with the soldiers.

My home was near Neosho, Newton County, Missouri, where I remained until I was about nine years old. My body was very feeble and it was a constant warfare between life and death to see who would gain the mastery.

From a child, I had an inordinate desire for knowledge, and especially music, painting, flowers, and the sciences, algebra being one of my favorite studies.

Day after day, I spent in the woods alone in order to collect my floral beauties, and put them in my little garden I had hidden in brush not far from the house, as it was considered foolishness in the neighborhood to waste time on flowers.

And many are the tears I had shed because I would break the roots or flowers of some of my pets while removing them from the ground, and strange to say all sorts of vegetation seemed to thrive under my touch until I was styled the plant doctor, and plants from all over the country would be brought to me for treatment. At this time I had never heard of botany and could scarcely read. Rocks had an equal fascination for me and many are the basketful that I have been compelled to remove from the outside chimney corner of the old log house, with the injunction to throw them down hill. I obeyed but picked up the choicest ones and hid them in another place, and somehow the same chimney corner would, in a few days or weeks, be running over again to suffer the same fate. I have some of the specimens in my collection now and consider them the choicest of the lot. Mr. and Mrs. Carver were very kind to me and I thank them so much for my home training. They encouraged me to secure knowledge helping me all they could, but this was quite limited. As we lived in the country, no colored schools were available. So I was permitted to go eight miles to a school at town (Neosho). This simply sharpened my appetite for more knowledge. I managed to secure my entire meager wardrobe from home, and when they heard from me I was cooking for a wealthy family in Ft. Scott, Kansas, for my board, clothes, and school privileges.

Of course, they were indignant and sent for me to come home at once to die, as the family doctor had told them I would never live to see 21 years of age. I trusted to God and pressed on (I had been a Christian since about eight years old). Sunshine and shadow were profusely intermingled such as naturally befall a defenseless orphan by those who wish to prey upon them.

My health began improving and I remained here for two or three years. From here to Olathe, Kansas, to school. From there to Paola Normal School. From there to Minneapolis, Kansas, where I remained in school about seven years finishing the high school, and in addition some Latin and Greek. From here to Kansas City, entered a business college of shorthand and typewriting. I was here to have a position in the union telegraph office as stenographer and typewriter, but the thirst for knowledge gained the mastery and I sought to enter Highland College at Highland, Kansas. Was refused on account of my color. I went from here to the Western part of Kansas where I saw the subject of my famous yucca and cactus painting that went to the World's Fair. I drifted from here to Winterset, Iowa, began as head cook in a large hotel. Many thanks here for the acquaintance of Mr. & Mrs. Dr. Milholland, who insisted upon me going to an art school, and chose Simpson College for me.

The opening of school found me at Simpson attempting to run a laundry for my support and batching to economize. For quite one month, I lived on prayer, beef suet, and cornmeal, and quite often being without the suet and meal. Modesty prevented me telling my condition to strangers.

The news soon spread that I did laundry work and really needed it, so from that time on favors not only rained but also poured upon me. I cannot speak too highly of the faculty, students, and in fact, the town generally. They all seemed to take pride in seeing if he or she might not do more for me than someone else.

But I wish to especially mention the names of Miss Etta M. Budd, my art teacher, Mrs. W.A. Liston & family and Rev. A.D. Field & family. Aside from their substantiate help at Simpson, were the means of my attendance at Ames.

I think you know my career at Ames and will fix it better than I. I will simply mention a few things. I received the prize offered for the best herbarium in cryptogram. I would like to have said more about you, Mrs. Liston & Miss Budd, but I feared you would not put it in about yourself, and I did not want one without all.

I received a letter from Mrs. Liston and she gave me an idea that it was not to be a book or anything of the kind this is only a fragmentary list.

I knit, crochet, and made all my hose, mittens, etc. while I was in school.

If this is not sufficient, please let me know, and if it ever comes out in print, I would like to see it.

God bless you all,
Geo. W. Carver

Photos I: Early Life

Above, Moses Carver House in GWC National Monument, Missouri. Below, Carver Memorial Cabin replica built by Henry Ford in Greenfield Village.

Clockwise from top left: as a student at Simpson College; as a young artist at Simpson College; as a Captain in the Iowa State Military.

II.

CLASSIFIED CHRONOLOGY OF CARVER'S LIFE AND LEGACY

A. Developmental Activities

1864   Birth date (estimated) of George Washington Carver (GWC) in Diamond Grove, Missouri

1885   Estimated date of completion of high school in Neosho, Missouri

1890   Enrolled in Simpson College to study art and music

1891   Transferred to Iowa State College in Ames

1893   Paintings entered in the Chicago Worlds Fair and received honorable mentions

1894   Earned Bachelor's Degree in Agriculture at Iowa State College

1894   Appointed member of faculty of Agriculture at Iowa State College

1896   Awarded the Master's Degree in Agriculture at Iowa State College

1896   Invited to Tuskegee Institute to head Department of Agriculture

1896   Appointed Director of the Experiment Station

1896   Made the initial farm and home demonstration visits in Macon County

1903 Requested by Booker T. Washington to design an appropriate way to deliver farm services and information to the people

1906 The first farm visits were made in the newly designed and constructed Jesup Wagon

1921 Appeared before the U.S. House of Representatives Committee on Ways and Means for Tariff on Peanuts

1935 Appointed collaborator for Mycology and Plant Disease Survey, Bureau of Plant Industry, USDA

1940 Established the Carver Research Foundation at Tuskegee Institute

1943 Died on January 5 and was buried in the Tuskegee Institute Cemetery near the grave of Booker T. Washington

1943 Entire estate bequeathed to Carver Research Foundation

B. Honorary Achievements

1928 Honorary Doctor of Science Degree, Simpson College

1939 Honorary membership, American Inventors Society

1941 Honorary Doctor of Science Degree, University of Rochester

1942 Honorary Doctor of Science Degree, Selma University

C. Citations and Awards

1916 Elected fellow of the Royal Society of the Arts, London, England

1923 Recipient of the Spingarn Award for distinguished service to science

1938 Featured in film made in Hollywood of the life of GWC

1939   Recipient of Roosevelt Medal for outstanding contributions to Southern agriculture

1941   Recipient of award of merit by Variety Clubs of America

D. Tributes By State Governments

1942   Official Marker authorized by the Governor of Missouri

1944   Governors of Connecticut, Illinois, Indiana, New Jersey, New York, Pennsylvania, and West Virginia issued proclamations designating the first week in January 1945 as George Washington Carver Week

E. Tributes by Colleges and Universities

1928   Awarded Honorary Doctor of Science Degree by Simpson College

1936   Bronze Bust unveiled at Tuskegee Institute as a tribute from friends throughout the nation for 40 years of creative research

1938   Tuskegee Institute Board of Trustees approved the creation and establishment of GWC Museum

1940   Tuskegee Institute established the Carver Research Foundation

1941   The Carver Museum dedicated at Tuskegee Institute by Henry Ford, Sr.

1941   Special exhibition of the GWC art collection at Tuskegee Institute

1943   Dr. Alma Illery initiates GWC Day Recognition

1955   North Carolina A&T University dedicated GWC Science Building on its campus

1956 Simpson College dedicated Science Building in memory of GWC

1968 Alabama A&M University dedicated and named GWC Complex

1970 Iowa State College established graduate scholarships in memory of GWC

1972 UMES GWC Science Building dedicated at Princess Ann, Maryland

1984 Tuskegee University dedicated GWC Experiment Station

1984 University of Missouri College of Agriculture established GWC Graduate Fellowships

1984 Dedication of GWC Room in the University of Maryland Memorial Union

1990 GWC Memorial Research Farm established by *Lincoln* University at Jefferson City, Missouri

F. Tributes by Industry

1939 Bronze sculpture dedicated to Spencer High School, Columbus, Georgia by Thom Houston Peanut Co.

1942 Erection of GWC Cabin, Greenfield Village, Dearborn, Michigan by the Ford Motor Company

1999 Erection of a monument along the walk of fame in Neosha, Missouri, by Carver District Birthplace Association and Benton Centennial Committee

G. Enshrinements

1943 Statue of Carver in his youth placed on site of Old Moses Carver Plantation of 210 acres which U.S. Congress designated a historic monument. Statue by Robert Amendola

1949   Statue by Christian Peterson, Artist in Residence, completed at Iowa State College. Located in the Carver Science Hall

1952   Carver was selected by Popular Mechanics Magazine as one of 50 outstanding Americans and listed in their 50th Anniversary Hall of Fame

1952   Selected by Popular Mechanics magazine as one of fifty (50) outstanding Americans and listed in their 50th Anniversary Hall of Fame

1977   Bronze bust enshrined in the New York Hall of Fame for Great Americans. Statue by Richmond Barthe

1986   Inducted into the Iowa Inventors Hall of Fame

1990   Inducted into the National Inventors Hall of Fame

1990   Inducted into the Alabama State Department of Archives and History

H. United States Patents

1925   Provision of a pomade or cream made from peanuts. Patent #1,522,176, dated 6/6/25

1925   Process for producing paints and stains from clays. Patent #1,541,478, dated 6/9/25

1927   Process of producing paints and stains. Patent #1,052,505, dated 6/14/27

I. Federal Recognition and Tributes

1921   Appearance before the U.S. House of Representatives, Committee on Ways and Means for Tariff on Peanuts

1943   78th Congress passed legislation HR641 P.L. 148 creating the GWC National Monument, Diamond Grove, Missouri

1943	George Washington Carver Liberty Ship designated and launched at Richmond, California, May 7
1946	79th Congress Joint Resolution P.L. 290, January 5 designated as GWC Day, issued by President Harry S. Truman
1947	U.S. postage stamp issued in honor of GWC
1948	First day of sale of GWC postage
1951	GWC fifty cents coin minted
1965	Tuskegee Institute designated by U.S. Congress as a National Historic Landmark
1965	GWC U.S. Polaris Submarine launched at New Port News, Virginia
1965	Carver plaque to hang permanently in Agriculture in recognition of scientist's great achievements. Presentation was made by Dr. Rosa L. Gragg, Chairman of the George Washington Carver Commemoration Committee; and Congressman Herbert Tenzer of New York
1974	U.S. Congress authorized the establishment of the Tuskegee Institute campus as a National Historic Site to include the campus as a Historic District, the Booker T. Washington Home, Grey Columns and the George Washington Carver Museum
1999	USDA named complex in honor of Dr. George Washington Carver

Photos II: Teaching

Carver and members of the faculty.

Carver at work in the classroom.

Carver and members of the faculty, about 1905.

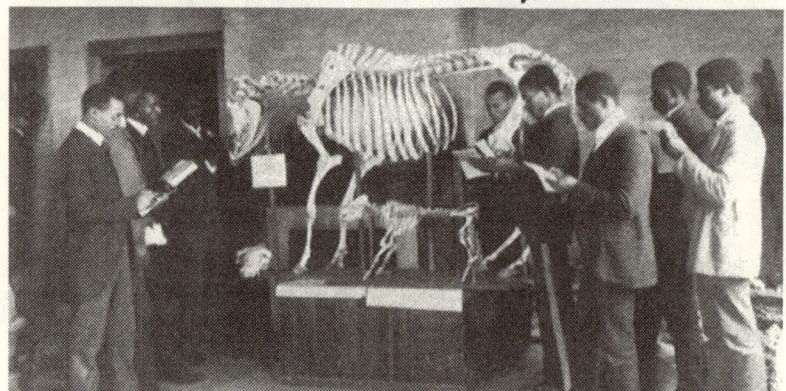
A class in anatomy.

Carver is honored.

III.

USDA COMPLEX NAMED IN CARVER'S HONOR

TU News Bureau Recognition

He was a genius. In addition to his unparalled contributions to agriculture, medicine and nutrition, this Tuskegee University scientist was an accomplished artist, painter and musician. He mastered the science of chemurgy, developing some 325 products from the peanut and more than 100 from the sweet potato and soybean.

His crop rotation theory is credited with revolutionizing southern agriculture. Soil depleted of much needed nutrients was revitalized by rotating crops other than cotton, thereby significantly increasing the productivity of all crops.

U. S. postage stamps bear the name of George Washington Carver, a U. S. ship was named in his honor; he has been inducted into the National Inventors' Hall of Fame, and he is recipient of the NAACP Spingarn Medal and the Theodore Roosevelt Medal.

Now, the U.S. Department of Agriculture has honored the contributions and achievements of Dr. Carver with the dedication Oct. 6 of the George. Washington Carver Complex in Beltsville, Maryland. Tuskegee University: President Dr. Benjamin F. Payton and the presidents of Simpson (Iowa) College and Iowa State University were among the participants in the dedication program.

"We are very pleased that the Federal Government has taken steps to honor one of the nation's outstanding scientists," Dr. Payton stated. "By naming a major, new federal complex in honor of George Washington Carver, the nation honors itself and lifts up for public recognition one of its most esteemed educators."

"Dr. Carver spent his entire life as a teacher and research scientist at Tuskegee University. His creative genius was so great that his impact was felt across the entire nation and in many places around the globe. Tuskegee University expresses its deep gratitude to all who played a part in making this significant naming opportunity a reality."

Born about 1864 in Diamond Grove, Mo., Carver taught himself to read and write. He was denied admission to Highland College because of his color, admitted to Simpson College where he studied piano and art; earned the bachelor's and master's degrees from Iowa State University and was subsequently Iowa's first African American professor; and joined the faculty at Tuskegee University at the invitation of Dr. Booker T. Washington.

It was during his 47 years at Tuskegee University, from 1896 until his death in 1943, that Dr. Carver impacted the scientific world. The peanut, which had not been recognized as a food crop when he arrived at Tuskegee University in 1896, was by 1940 second only to cotton as a cash crop in the south.

The George Washington Carver Center named in his honor is a sprawling 350,000 square foot facility on a 45-acre campus. In addition to the office space it is now providing four government agencies and their 1,000 employees, the GWC Center also includes: a cafeteria that will serve breakfast and lunch, a fitness center with provisions for weight lifting and aerobics; a health center with emergency bed space; a credit union; and a day care center that will accommodate at least 45 children.

The Carver Garden is being landscaped so that children attending the day care center will be able to learn about Dr. Carver's work with plants.

The U. S. Department of Agriculture-sponsored George Washington Carver Recognition Week activities got under way Oct. 4 when Agriculture Secretary Dan Glickman presented Carver coloring books and soybean crayons to children from Van Ness and Lincoln Multicultural schools.

A symposium during the week of Recognition Activities will include presenters from Tuskegee University, Iowa State University, the Director/Curator of the George Washington Carver National Monument, and Peter Burchard, author of Carver: A Great Soul, Will provide an overview of Dr. Carver's life. Dr. B. D. Mayberry of Tuskegee University is a former student of Dr. Carver, Dr. Sande McNabb of Iowa State University will talk about how his career path in plant pathology was influenced by the work of Dr. Carver, and Bill Jackson is the Curator/Director of the GWC National Monument at Diamond Grove, Mo., the birthplace of Dr. Carver.

Children from schools in Maryland, Virginia, and Washington, D.C., were invited to view exhibits at the GWC Center Oct. 6-7. The exhibits, all celebrating Dr. Carver's life and work, were developed by Simpson College, Iowa State University, and Tuskegee University

As a part of its exhibit, Tuskegee University presented an interactive opportunity for visitors to learn about Dr. Carver, the University's current research in biotechnology and hydroponic crop growth, both using the sweet potato and peanut, was featured, and foods made by Tuskegee University researchers from sweetpotatoes were displayed, some available for tasting.

IV.

CARVER NAMESAKES BY STATE

Alabama
1. Carver Elementary School, 2001 West Fairview Ave., Montgomery, AL 36108-4118, 205-269-3625

2. Carver Junior High, 2001 W. Fairview Ave., Montgomery, AL 36108-4118, 205-269-3640

3. Carver Senior High School, 2001 W. Fairview Avenue, Montgomery, AL 36108, 205-269-3636

4. Carver Creative Performing Arts Center, 2001 W. Fairview Ave., Montgomery, AL 36108, 205-269-3813

5. Carver Day Nursery, 1109 Jacksonville Ct., Gadsden, AL 35901, 205-547-4325

6. Carver Elementary School, 527 Greensboro St., Eutaw, AL 35462, 205-552-3052

10. Carver High School, 3400 33rd Terrace N., Birmingham, AL 35207, 205-849-3500

11. Carver Middle School, Flomaton, AL 32535, 904-256-3788

12. Carver Primary School, 307 Carver Ave., Opelika, AL 36801, 205-745-9712

13. Carver State Technical College, 414 Stanton, Mobile, AL 36617, 205-473-8692

14. The Carver Complex, Alabama A&M University, Normal, AL 35762, 205-851-5783

15. Carver Middle School, 801 E. North Street, Dothan, AL 36303, 205-794-1440

16. Carver Jr. High School, 527 Greensboro Street, Eutaw, AL 35462, 205-372-4816

17. The George Washington Carver Museum, The National Park Service, Tuskegee Institute, AL 36088, 205-727-3200

18. The Carver Research Foundation of Tuskegee University, Tuskegee Institute, AL 36088, 205-727-8961

19. The GWC Experiment Station, Tuskegee University, Tuskegee Institute, AL 36088, 205-727-8333

20. George Washington Carver Subdivision, Carver Court, Tuskegee Institute, AL 36088, dedicated in 1943

21. Carver Super Service Station, 1703 W. Montgomery Road, Tuskegee Institute, AL, 36088, 205-727-2250

Arkansas
1. Carver Elementary School, 309 North Linden, Pine Bluff, AR 71601-3423, 501-534-4813

2. George Washington Carver, 1116 West 14th Street, Little Rock, AR 72202

Arizona
1. Carver Elementary School, 1300 W. 5th St., Yuma, AZ 85364-2899, 602-782-1843

California

1. George Washington Carver Elementary School, 1360 Oakdale Avenue, San Francisco, CA 94124-2724, 415-822-6391

2. Carver Elementary School, 2463 S. Fig Ave., Fresno, CA 93706-4999, 209-441-3058

3. George Washington Carver School, 5335 East Pavo St., Long Beach, CA 90808-3599, 213-420-2697

4. Carver Elementary School, 3251 Juanita St., San Diego, CA 92105-3807, 619-293-8325

5. Carver Elementary School, 11150 Santa Rosalia St., Stanton, CA 90680-3193, 714-663-6437

6. Carver Elementary School, 1300 San Gabriel Blvd., San Marino, CA 91108-2799

7. George Washington Carver Elementary, 1425 E. 120th St., Los Angeles, CA 90059-2499, 213-898-6150

Colorado

1. Carver Elementary School, 4740 Artistic Circle, Colorado Springs, CO 80917-2199, 719-520-2225

2. George Washington Carver, 2270 Humboldt Street, Denver, CO 80205-5330, 303-861-4588

District of Columbia

1. Carver Elementary School, 45th & Lee Streets NE, Washington, DC 20019-3800, 202-724-4602

Florida

1. Carver Early Learning Center, 1142 Laurel Street,, Tampa, FL 33607-5599, 813-251-4192

2. George Washington Carver School, 2854 West 45th Street,, Jacksonville, FL 32208, 204-765-1323

3. Carver Elementary School, 238 Grand Ave.,, Miami, FL 33133-4897, 305-443-5286

4. GWC Jr & Sr High School, 4901 Lincoln Road, Coral Gables, FL 305-444-7388

5. GWC Middle, Junior & Senior High School, 4901 Lincoln Drive, Cocanut Grove Station, Miami, FL 33133, 305-444-7388

Georgia

1. Carver Elementary School, P.O. Box 658,, Dawson, GA 31742-0658, 912-995-5451

2. Carver Elementary School, East Walton St.,, Milledgeville, GA 31061-0000 912-452-1996

3. Carver Elementary School, 3042 8th Street., Columbus, GA, 31900, 404-323-6347

4. The Carver State Bank, P.O. Box 2769, Savannah, GA 31402, 912-233-9971

Illinois

1. George Washington Carver DCC, 710 West Third St.,, Peoria, IL 61605-2299, 309 674-2915

2. George Washington Carver School, P.O. Box AE,, Hopkins Park, EL 60944, 815-944-5069

Indiana

1. Carver Day Care School, 100 E. Walnut,, Evansville, IN 47713-1999, 812-423-2683

2. Carver Elementary School, 2535 Virginia Street.,, Gary, IN 46407-3727, 219-886-6545

Iowa

1. Carver Science Building, Simpson College, Indianola, Iowa 50125

2. George Washington Carver Hall, Iowa State University, Ames, Iowa 50010

Louisiana

1. Carver Elementary *School,* Bastrop, LA 71220, 318-281-3832

2. Carver Elementary *School,* 1300 Orange, Monroe, LA 71201, 318-322-4245

3. Carver Branch Public Library, 2941 Renwick, Monroe, LA 71201, 318-323-3129

4. George Washington Carver Elementary, Martin Luther King Drive, DeRidder, LA 70634, 318-463-7380

5. Carver Federal Credit Union, 1544 Milam, Shreveport, LA 71100, 318-222-3626

6. Carver Elementary School, Luling, LA 70070, 504-763-6626

7. George Washington Carver Middle School, 3019 Higgins Blvd., New Orleans, LA 70130, 504-942-1796

8. George Washington Carver Senior High School, 3059 Higgins Blvd., New Orleans, LA 70130, 504-942-1775

9. Carver Branch library, 1214 East Boulevard, Raton Rouge, LA 70802-6521, 504-389-4978

10. Carver Recreational Facility, 838 South Peters Street, New Orleans, LA 70130

Maryland
1. George Washington Carver Vocational Technical High School, Bentalaw & Preston Streets, Baltimore, MD 21200

2. George Washington Carver Science Auditorium, University of Maryland, Eastern Shore,, Princess Anne, MD, 21853

Mississippi
1. Carver Elementary School, Jefferson, Indianola, MS 38751, 601-887-2353

2. Carver Junior High School, Raymond, MS 39154, 601-857-5006

3. Carver School, 910 N. Green, Tupelo, MS 38801, 601-841-8875

4. Carver Junior High School, 900 44th Avenue, Meridian, MS 39301, 601-484-4482

5. Carver Middle School, Jefferson, Indianola, MS 38751, 601-887-1942

Missouri
1. Carver Christian School, 5000 East 17th St., Kansas City, MO 64127-2833, 816-231-2866

2. GWC Room, Memorial Union Building, University of Missouri, Columbia, MO 65211 314-882-8308

3. GWC Research Farm, Lincoln University, Jefferson City, MO 65191, 314-681-6120

4. GWC Memorial, 906 Westminister Avenue, Fulton, MO 65251, 314-642-5551

5. GWC National Monument, Box 38, Diamond, MO 64840, 417-325-4151

New York
1. Carver Federal Savings Bank, New York, NY

North Carolina
1. George Washington Carver Agricultural Building, NC A&T University, Greensboro, NC 27411, 919-334-7979
2. Carver High School, 3545 Carver Road, Winston Salem, NC 27105, 919-727-2987

Ohio
1. Carver Community Center, 165 W. 4th St.,, Chillicothe, OH 45601-3210, 614-773-4242
2. George Washington Carver Elementary, 2201 East 49th Street,, Cleveland, OH 44103-4403, 216-391-2916

Pennsylvania
1. George Washington Carver High School for Engineering and Science, Philadelphia, PA

South Carolina
1. Carver Child Development Center, 2100 Waverly St., Columbia, SC 29204, 605-343-2900
2. Carver Child Development Center, 1001, W. Sumter St., Florence, SC 29501, 605-664-8156

Tennessee

1. Carver High School, 1591 Pennsylvania St., Memphis, TN 38109, 901-775-7594

Texas

1. Carver Academy School, 1905 NW 12th Street, Amarillo, TX 79107, 806-371-5630
2. Carver Academy, 2302 Holland, Marshall, TX 75670-5956, 214-935-3916
3. Carver 6th Grade Center School, Mailing: P.O. Box 27 Waco,, TX 76703-0027, Location: 1601 Dripping Springs Rd., Waco, TX 76704, 817-752-2526, 816-757-0787
4. George Washington Carver Elementary School, 603 Lakeway Dr., Georgetown, TX 78628-2843, 512-863-3135
5. Carver Branch Library, 3350 E. Commerce, San Antonio,, TX 78220-1220, 512-225-7801
6. Carver Elementary School, 1401 Martin Luther King, Bryan, TX 77803-5712
7. Carver Center, P.O. Box 3912, Odessa, TX 79760, 915-337-6481
8. The Carver Cultural Center, 226 N. Hackberry St., San Antonio, TX 78220
9. The Carver Institute, 2646 South Loop, W., Suite 275, Houston, TX 77054-2640

Virginia
1. Carver Elementary School, 1110 West Leigh Street, Richmond, VA 23230, 804-780-6247

2. George Washington Carver Middle School, Chesterfield School Board, Chester, Virginia

3. Carver Magnet School for Math and Science, Richmond, VA

West Virginia
1. Carver Career & Technical Education Center, 4799 Midland Drive, Charleston, WV 25306-6397, 304-348-1965

Photos III: Research and Experiments

The Jesup wagon.

Above: Specimens in the Carver Museum.

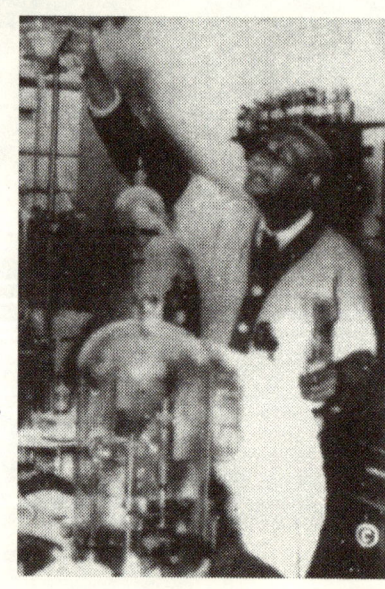

Above: Dr. Carver and Curtis.
Right: Dr. Carver at work.

Dr. Carver at work, measuring chemicals.

Carver at work in the laboratory.

V.

CARVER'S ACHIEVEMENTS

A. Products made from sweet potatoes*
*Figure in parentheses indicates number of products of that type

Foods
After Dinner Mints (3)
Bisque Powder
Breakfast Food (5)
Candies (14)
Chocolate
Coffee, dry
Dried Potatoes (2)
Dry Paste
Egg Yolk
Granulated Potatoes
Flour (4)
Instant Coffee
Lemon Drops
Meal (4)
Mock Coconut
Molasses (3)
Orange Drops
Potato Nibs
Sauce

Spiced Vinegar
Starch
Sugar
Synthetic Ginger
Tapioca
Vinegar
Yeast

Stock Feeds
Hog Feed
Stock Feed Meal (3)

General
Alcohol
Dyes
Fillers for Wood (14)
Library Paste
Medicine
Paints
Paper (from vines)
Rubber Compound
Shoe Blacking
Synthetic Cotton
Stains
Synthetic Silk
Writing Ink

B. Products Made from Peanuts
(As Compiled by the Carver Museum)
Beverages
Beverage for Ice Cream
Blackberry Punch
Evaporated Peanut Beverage
Cherry Punch

Normal Peanut Beverage
Peanut Beverage Flakes
Peanut Lemon Punch
Peanut Kuomiss Beverage
Peanut Orange Punch #1
Peanut Punch #2
Plum Punch

Cosmetics
All Purpose Cream
Antiseptic Soap
Baby Massage Cream
Face Bleach and Tan Remover
Face Crem
Face Lotion
Facte Ointment
Face Powder
Grease
Fat Producing Cream
Glycerine
Hand Lotion
Oil for Hir and Scalp
Peanut Oil shampoo
Pomade for Scalp
Pomade for Skin
Shampoo
Shaving Cream
Totter and Dandruff Cure
Toilet Soap
Vanishing Cream

Dyes, Paints, and Stains
Dyes for Cloth (30)
Dyes for Leather (19)

Paints
Wood Stains (17)
Special Peanut Dye

Stock Foods
Hen Food for Laying (peanut hearts)
Molasses Feed
Peanut Hay Meal
Peanut Hull Meal
Peanut Meal
Peanut Stock Food (3)

Foods
Bar Candy
Breakfast Food (5)
Bisque Powder
Butter from Peanut Milk
Caramel
Cheese Cream
Cheese Nut Sage
Cheese Pimento
Cheese Tutti Frutti
Chil Sa
Choose Sandwich
Chocolate Coated Peanuts
Chop Suey Sou
Cocoa
Cooking Oil
Cream Candy
Cream from Milk
Crystallized Peanuts
Curds
Dehydrated Milk Flair
Dry Coffee

Flavoring Paste
Golden Nuts
Instant Coffee
Lard Compound
Malted Substitutes
Mayonnaise
Meat Substitutes
Milks (32)
Mock Goose
Mock Chicken
Mock Meat
Mock Veal Cutlet
Mock Oyster
Oleomargarine
Pancake Flour
Peanut Bar #1
Peanut Bisque Flour
Peanut Brittle
Peanut Butter, regular (3)
Peanut Cake (2)
Peanut Chocolate Fudge
Peanut Dainties
Peanut Flakes
Peanut Flour (11)
Peanut Hearts
Peanut Kisses
Peanut Meal, brown
Peanut and Popcorn Bars
Peanut Relish (2)
Peanut Sausage
Peanut Surprise
Peanut Tofu Sauce
Peanut Wafers
Pickle, plain

Salted Peanuts
Shredded Peanuts
Substitute Asparagus
Sweet Pickle
Vinegar
White Pepper, from Vines
Worcestershire Sauce

Medicines
Castoria Substitute
Emulsion for Bronchitis
Goiter Treatment
Iran Tonic
Laxatives
Medicine similar to Castor Oil
Oils, Emulsified with Baby
Mercury for Venereal Disease (2)
Rubbing Oil
Tannic Acid
Quinine

General
Axle
Charcoal from Shells
Cleaner for Hands
Coke (from hull)
Diesel Fuel
Fuel Briquettes
Gas
Gasoline
Illuminating Oil
Insecticide
Insulating Boards (18)
Laundry Soap

Linoleum
Lubricating Oil
Nitroglycerine
Paper (colored) from skins
Paper (Kraft) from vines
Paper (white) from vines
Printer's Ink
Plastics
Rubber
saw oil
Shoe and Leather Blacking
Sizing for Walls
Soap Stock
Soil Conditioner
Sweeping Compound
Wall Boards (from hulls) (11)
Washing Powder
Wood Filler

C. General Products from Other Materials
Alcohol
Axle Grease
Charcoal from Shells
Cleaner for Hands
Coke (from hull)
Diesel Fuel
Dyes (73)
Fillers for Wood (14)
Fuel Briquettes
Gas
Gasoline
Glue
Illuminating Oil
Insecticide

Insulating Boards (18)
Laundry Soap
Library Paste
Linoleum
Lubricating Oil
Medicine
Nitroglycerine
Paints
Paper (white) from vines
Paper Kraft from vines
Paper (colored) from skins
Plastics
Printer's Ink
Rubber
Shoe and Leather Blacking
Sizing for Walls
Soap Stock
Soil Conditioner
Sweeping Compound
Synthetic Cotton
Synthetic Silk
Wall Boards (from hulls) (11)

D. Publications & Bulletins by George Washington Carver

"Plants as Modified by Man, " B.A. thesis, Iowa State College, 1894.
   "Grafting the Cacti," *Report of the Iowa State Horticultural Society,* 1893, pp. 257-259.
"Best Ferns for the North and Northwest," *Iowa Agricultural College Experiment Station Bulletin,* No. 27 (1895), pp. 150-153.
"Treatment of Currants and Cherries to Prevent Spot Diseases," *Iowa Agricultural College Experiment Station Bulletin,* No. 30 (1895), pp. 289-301.
"Fungus Diseases of Plants at Ames, Iowa, 1895," *Proceedings of the*

Iowa Academy of Sciences. Vol. 3, (1895), pp. 140-148. By George Washington Carver and L. H. Pammel.

"Inoculation Experiments with Gymnosporan Gium Macropus LK," *Proceedings of the, Iowa Academy of Sciences,* Vol. 3 (1895), pp. 162-169. By George Washington Carver and F. C. Stewart.

"Our Window Gardens," *Iowa Agricultural College Experiment Station Bulletin,* No. 32 (1896), pp. 516-525.

"Many Food Products Can Be Made From Peanut and Sweet Potato," *The American Food Journal,* Vol. 16, No. 8 (August, 1921), p. 20.

Forty-four bulletins and seven circulars were published by Tuskegee Institute from 1898 to 1943. The bulletins contain information on soil building, crop rotation, property improvements, use of natural fertilizers, utilization of local and natural resources and expanded industrial use of farm products. The circulars deal with farming and the home use of agricultural products. A list of these publications follows.

E. Farm and Home Bulletins by George Washington Carver

NO.	DATE	TITLE
1	1898	Feeding Acorns
1	1940	The Farmer's Almanac**
		**Originally this was not a bulletin. When revised in 1940, it was made a bulletin with this number as No. 1 was no longer in use.
2	1898	Experiments with Sweet Potatoes
3	1899	Fertilizer Experiment with Cotton
4	1901	Some Cercospora of Macon Co., Alabama
5	1903	Cow Peas
6	1905	Cotton Growing on Sandy Upland Soils
7	1905	How to Build up Worn-Out Soils
8	1906	Successful Yields of Small Gram
9	1906	The San Jose Scale in Alabama

10	1906	Saving the Sweet Potato Crop
11	1908	Relations of Weather and Sod Conditions to the Fruit Industry of South-east Alabama
12	1907	Saving the Wild Plum Crop
13	1908	How to Cook Cow Peas
14	1908	How to Make Cotton Growing Pay
15	1909	Increasing the Yield of Corn
16	1909	Some Ornamental Plants of Macon Co., Alabama
17	1910	Possibilities of the Sweet Potato in Macon County
18	1910	Nature Study and. Gardening for Rural Schools
19	1911	Some Possibilities of the Cow Pea in Macon County
20	1911	Cotton Growing for Rural Schools
21	1911	White and Colored Washing with Native Clays from Macon County, Alabama
22	1912	Dairying in Connection with Farming
23	1912	Poultry Raising in Macon County,
24	1912	The Pickling and Curing of Meat in Hot Weather
25	1913	A Study of the Soils of Macon Co., Alabama and their Adaptability of Certain Crops
26	1915	A New and Prolific Variety of Cotton
27	1915	%%en, Mix and How to Can and Preserve Fruits and Vegetables in the Home
28	1915	Smudging an Orchard with Native Material in Alabama
29	1915	Alfalfa, the King of all Fodder Plants. Successfully Grown in Macon County

30	1915	Possibilities of the Sweet Potato in Macon County (Revision of #17)
31	1916	How to Grow the Peanut and 105 Ways of Preparing it for Human Consumption
32	1916	Three Delicious Meals Every Day for the Farmer
33	1917	Twelve Ways to Meet the New Economic Conditions here in the South
34	1917	Forty-three Ways to Save the Wild Plum Crop
35	1917	How to Grow the Cow Pea and 40 Ways to Prepare it as a Table Delicacy
36	1918	How to Grow the Tomato and 105 Ways to Prepare it for the Table
37	1918	How to Make Sweet Potato Flour, Starch, Bread, Sugar and Mock Cocoanut
38	1918	How the Farmer Can Save His Sweet Potatoes'
39	1927	How to Make and Save Money on the Farm
40	1935	The Raising of Hogs
41	1936	Can Live Stock Be Raised Profitably in Alabama?
42	1936	How to Build Up and Maintain the Virgin Fertility of Our Soils
43	1942	Nature's Garden for Victory and Peace
44	1943	The Peanut

(Below: C = Circular, L= Leaflet)

C	1912	The Canning and Preserving of Fruits and Vegetables in the Home
L	1915	A New and Prolific Variety of Cotton
L	1915	How to Raise Pigs with Little Money
L	1916	How to Live Comfortably this Winter

L	1916	What Shall We Do for Fertilizer this Year?
L	1931	Some Peanut Diseases
L	1938	Some Choice Wild Vegetables That Make Fine Foods
L	1938	How to Dry Fruits and Vegetables

Other Manuscripts

1931	Some Peanut Disease
1938	Some Choice Wild Vegetables (Special Leaflet #1)
1938	A Resume of the Fungus Collections of George Washington Carver
1938	How to Dry Fruits and Vegetables

F. Quotations of George Washington Carver

"A friend is a priceless possession. Material wealth may disappear as rapidly as the tiny snow flake just as a block of ice in the fierce heat of a tropical sun. A true friend is a priceless heritage. Eternity alone can measure their value."

"Race and creed find no recognition in the eyes of the Deity when He bestows His generous gifts."

"Divine love is destined to rule the world, I believe, despite the many things that often irritate and depress us. "

"All good teaching is scientific, because it is simply the truth, and the simplest and clearest way it can be presented to the pupil constitutes good teaching."

"How far you go in life depends on your being tender with the young, compassionate with the aged, sympathetic with the striving, and tolerant of the weak and the strong. Because someday in life you will have been all of these."

"My dear friend, watch how you give advice. Wise men don't need it, fools won't heed it."

"It would be a good idea if more of us would listen more and talk less. No one ever learns anything by talking."

"No one will ever get out of this world alive."

"To me nature in its varied forms are the little windows through which God permits me to commune with Him, and to see much of His glory, by simply lifting the curtain, and looking in. I love to think of nature as wireless telegraph stations through which God speaks to us every day, every hour, and every moment of our lives."

"Never get in the habit of making excuses. They don't get you anywhere. Ninety-nine percent of the failures come from people who have the habit of making excuses."

"It is not the style of clothes one wears, neither the kind of automobile one drives, nor the amount of money one has in the bank, that counts. These mean nothing. It is simply service that measures success."

"Without God to draw aside the curtains I would be helpless. God did nothing without reasons."

"Start where you are with what you have. Make something of it; never be satisfied!"

"I have only tried to do my duty to God, my country, and race -as rapidly as the Great Creator gives me light and strength."

"Let every occasion be a great occasion, for you cannot tell when fate may be taking your measure for a larger place."

"Look about you. Take hold of the things that are here. Let them talk to you. You learn to talk to them.

"Learn to do common things uncommonly well. We must always keep in mind that anything that helps fill a dinner pail is valuable."

"There is no short cut to achievement. Life requires thorough preparation-veneer isn't worth anything."

"One reason I never patent my products is that if I did it would take so much time I would get nothing else done. But mainly I don't want my discoveries to benefit specific favored persons."

"I love to think of nature as an unlimited broadcasting system, through which God speaks to us every hour, if we will only tune in."

"As the sunset of life draws nearer and nearer to its setting, my anxiety becomes more acute for the preservation of my life's work. I am hoping it can go on and on, blessing unborn generations."

G. ONE OF GEORGE WASHINGTON CARVER'S FAVORITE POEMS

The poem "Equipment" by Edgar Guest was one of Dr. Carver's favorites. He often read it to students as inspiration.

EQUIPMENT

Figure it out for yourself, my lad,
You've all that the greatest of men have had;
Two arems, two hands, two legs, two eyes;
And a brain to use if you would be wise,
With this equipment they all began.
So start for the top and say "I can."

A CARVER HANDBOOK

Look them over, the wise and great,
They take their food from a common plate
And similar laces they tie their shoes.
The world considers them brave and smart,
but you've they had when they made their start.

You can triumph and come to skill;
You can be great if you only will.
You're well equipped for what fight you choose;
You have legs and arms and a brain to use,
And the man who has risen great deeds to do
Began his life with no more than you.

You are the handicap you must face,
You are the one who must choose your place,
You must say where you want to go,
How much you will study the truth to know;
Good has equipped you for life, bu He
Lets you decide what you want to be.
Courage must come from the soul within
The man must furnish the will to win,
So figure it out for yourself, my lad
You were born with all that the great have had,
With your equipment they all began,
Get hold of yourself and say: "I can."

—Edgar A. Guest

VI.

COMPREHENSIVE CHRONOLOGY

A. Carver's Life 1864-1943

1864   Carver was born on the farm of Moses and Sarah Carver at Diamond Grove, Missouri, where his mother was a slave. The identity of his father is unknown, but he is believed to have been a slave on a neighboring farm who died shortly after young Carver's birth.

1865   In the waning days of the Civil War, mother and baby were abducted by slave raiders. George was later recovered by Moses Carver in exchange f or a three hundred dollar race horse, but his mother was never found.

1866   After the slaves were freed, the carvers reared George and his older brother, Jim an members of their family. Too frail and sickly to do heavy farm work, George learned instead to cook, sew and launder. Much of his free time, however, was spent out in the woods collecting wild flowers, stones and insects.

1874   Young George learned to read at an early age. Eager for a formal education, he left the farm with the Carvers' blessing to attend a school for black children established by the Freedmen's Bureau in Neosho, Missouri.

1876   In search of further schooling, Carver joined the westward migration to Kansas. For several years he worked at odd jobs at Fort Scott, attending school whenever the opportunity arose.

In 1879, however, he fled Fort Scott in horror after witnessing the lynching of a black man.

1881   After working for two years as a laundryman and itinerant field hand in several Kansas towns, Carver moved to Minneapolis, Kansas, to complete his high school education. Upon graduating he applied to the Presbyterian College at Highland, Kansas, and was accepted by mail. When he arrived at Highland, however, he was denied admission because of his race.

1882   His hopes for a college education dashed for the moment, Carver became a homesteader in Ness County, Kansas. He built his own sod house and struggled to raise chickens and vegetables on arid land. While living in Ness County, he was elected assistant editor of the local literary society's newspaper and began to demonstrate a talent for painting.

1885   Carver estimated date of completed high school in Neosho, Missouri

1890   Restless and discouraged by his fruitless efforts at homesteading, Carver pulled up stakes and made his way to Iowa. At Winterset, Iowa, he was befriended by a couple, the Milhollands, who were impressed by his artistic ability and encouraged him to seek training. He applied to Simpson College in Indianola and was admitted. After paying his fees, Carver was left with only ten cents, whereupon he opened a laundry that supported him through college. Although Carver had come to Simpson to study art, his instructors, aware of his love for nature and concerned about his ability to earn a living as an artist, urged him instead to pursue a degree in science.

1891   Carver transferred to Iowa State College of Agriculture and Mechanical Arts at Ames. Active in student affairs, he became the college football team's trainer.

1893   Carver's paintings were exhibited in Chicago.

1894 Carver Received the Bachelor's Degree in Agriculture from Iowa State College.

Carver appointed to the faculty of Iowa State college as an assistant in botany while studying for a master's degree. He published two papers of scientific merit in the field of mycology (the study of plant fungi), an interest he was to maintain throughout his life. The first black graduate of the college, Carver was also its first black faculty member and the first black person to receive an advanced degree from Iowa State.

1896 Carver received the Master's Degree from Iowa State College.

Booker T. Washington, principal of Tuskegee Institute in Macon County, Alabama, asked Carver to direct the school's agriculture department and experiment station. While he would have preferred to devote himself to the pure science for which he had been trained, Carver realized that the opportunity for service to his people would be greater at Tuskegee.

1897 Carver was appointed Director of the Tuskegee Experiment Station which had been approved by the State Legislature and given an annual appropriation of $1500.00.

Demonstrating his lifelong preoccupation with salvaging waste materials, Carver improvised a laboratory at Tuskegee out of junk-pile scrap. He also set up an experimental farm, where he quickly revitalized the exhausted soil and improved crop yields dramatically. Concerned about the strangle-hold that the cotton single-crop system still maintained on the Southern economy, he experimented with such crops as cowpeas, soybeans, peanuts, and sweet potatoes. In 1898 he published the first of 32 bulletins intended to teach modern agricultural techniques to impoverished local farmers. He still found time to paint, however, and he occasionally lectured on art to the student body.

1903 Carver was requested by Booker T. Washington to design and construct a more appropriate wagon to use for farm visits.

1906   The newly designed and constructed Jesup Wagon made its first movable school farm visit.

1916   Consistent advocate of crop diversification as a solution to the South's dependence on cotton, Carver sought to demonstrate the versatility of new crops. He issued a bulletin on growing and preparing peanuts for human consumption and began research on new uses for the plant and its by-products. Those uses, which in time would number 300, included wallboard, cosmetics, cough syrup and stains.

Elected fellow of the Royal Society of the Arts, London, England

At the same time he experimented with the sweet potato as both a foodstuff and an ingredient in synthetic compounds.

Carver was elected, a Fellow of the prestigious Royal Society of Arts of London. 1

1920   Carver was invited to speak before the convention of the United Peanut Association about the myriad of uses he had developed for the peanut.

1921   Impressed by Carver's presentation, the peanut growers asked him to testify before the Ways and Means Committee of the U. S. House of Representatives in support of a protective tariff for their commodity. His successful appearance before the committee drew national attention and fixed his identification with the peanut firmly in the public mind.

1922   Concerned about lynchings, violence and racial injustice, Carver became active in the promotion of interracial harmony. He toured throughout the South for nearly 20 years. Beginning in 1916 he appeared under the sponsorship of the YMCA. Later he served with the Commission on Interracial Cooperation, a group concerned with encouraging understanding between the races and ameliorating the conditions under which black Southerners lived.

1923   Carver began to receive numerous awards and honors. The National Association for the Advancement of Colored People gave him the Spingarn Medal for distinguished research. Following commendations from the United Daughters of the Confederacy and the Royal Geographic Society of London, Carver modestly made light of his talents and attributed his successes to the Lord.

Recipient of the Spingarn Award for distinguished service to science.

1925   Advancing age, troubled health, and a desire to further investigate some of his ideas persuaded Carver to give up teaching in favor of full-time research. Earlier, a group of businessmen had formed the Carver Products Company with no great success. For a time Carver's copyrighted cough medicine was marketed through the Carver Penol Company, though with little return to its inventor or the investors. Carver's research during this period focused primarily on peanut by-products, but he also experimented with industrial uses for sweet potatoes, cotton, corn, and petroleum.

Carver was awarded a U.S. patent for development of a pomade or cream made from peanuts. Patent 11,522,176, dated June 6, 1925   Carver was awarded a U.S. patent for a process for producing paints and stains from clays. Patent 11,541,478, dated June 9, 1925.

1927   Carver was &warded a U.S. patent for a process of producing paints and stains. Patent #1,052,505, dated June 14, 1927.

1928   Carver received the Honorary Degree of Doctor of Science from Simpson College in Indianola, Iowa.

1931   Carver undertook lecture tours of the Midwest, the Northeast and the upper South. He wrote that he was "being invited to places to speak where we thought a few years ago, that would never open up to colored people." Growing national prominence brought a deluge of requests to speak, but recurrent illness forced him to decline most invitations.

1932   For the next seven years, Carver was to spend much of his time exploring the benefits of peanut oil massage in the treatment of a number of afflictions, particularly muscle damage resulting from poliomyelitis. Though he gained much publicity for discovering a "cure" for polio, Carver vehemently denied any such claim. Nevertheless, thousands came to Tuskegee for therapeutic massage, and many believed themselves helped by the treatment. Carver sought, but never won, the endorsement of the American Medical Association for his method.

1935   The U.S. Department of Agriculture appointed Carver a collaborator in its Plant Disease Survey in recognition of the work with plant fungi over the years. Several native American fungi are named for him, since he was the first to identify them.

1936   Bronze Bust of Carver was unveiled at Tuskegee Institute on June 2, a tribute from his friends throughout the nation for his 40 years of creative research

1938   Feature Film, "Life of George Washington Carver," was made in Hollywood by the Pete Smith Specialty Company.

The George Washington Carver Museum was established by Board of Trustees of Tuskegee Institute.

1939   Bronze sculpture dedicated to Spencer High School, Columbus, Georgia by Thom Houston Peanut Co.

Carver was recipient of Roosevelt Medal for outstanding Contribution to Southern Agriculture.

Carver elected honorary membership, American Inventors Society.

1940 In order that his work might continue after his death, Carver established the George Washington Carver Research foundation at Tuskegee Institute. He bequeathed his life savings to the foundation.

1941 The George Washington Carver Museum was founded by Tuskegee Institute to display his scientific and artistic accomplishments, his honors and awards. the museum was designated a national treasure in 1976 when it became a main feature of the Tuskegee Institute National Historic Site. The Museum was dedicated by Henry Ford, Sr. on March 11.

Carver was awarded Honorary Doctor of Science Degree, University of Rochester.

Carver received Award of Merit from Variety Clubs of America.

1942 Carver received Honorary Doctor of Science Degree from Selma University, Selma, AL.

The George Washington Carver Cabin was erected by Henry Ford to honor and commemorate Dr. Carver's achievements and contributions to American life, Greenfield Village at Dearborn, MI.

Official marker was authorized by the Governor of Missouri.

Erection of GWC Cabin, Greenfield Village, Dearborn, Michigan by the Ford Motor Company.

1943 George Washington Carver died at Tuskegee Institute, AL on January 5.

B. Carver's Legacy, 1943 to 1990

1943   Carver estate amounting to over $60,000 bequeathed to the George Washington Carver Foundation.

78th Congress passed legislation H.R. 647, Public Law 148, creating the GWC National Monument, Diamond Grove, Misssouri. Legislation sponsored by Representative William Short and Senator Harry S Truman of Missouri.

Statue of Carver in his youth placed on site of Old Moses Carver Plantation of 210 acres which U.S. Congress designated a history monument. Statue by Robert Amendola.

GWC Liberty Ship was designated and launched at Richmond, CA on May 7.

1944   Governors of Connecticut, Illinois, Indiana, Now Jersey, New York, Pennsylvania, and West Virginia issued proclamations designating the first week in January 1945 an George Washington Carver Week.

1946   79th Congress Joint Resolution, Public Law 290, January 5, designated as George Washington Carver Day, issued by President Harry S. Truman, in response to a campaign by Dr. Alma Illary of Pittsburgh, Pennsylvania.

1947   U.S. Postage stamp in honor of Carver was issued.

Carver Museum damaged by fire (restored 1951).

1948   First day sale of the Three-cent carver Commemorative Stamp.

1949   Statue of Carver completed at Iowa State College by Christian Peterson, Artist in Residence.

1951   Fifty-cent piece was coined to likeness of Carver and Booker T. Washington.

1952 Carver was selected by Popular Mechanics Magazine as one of 50 outstanding Americans and listed in their 50th Anniversary Hall of Fame.

1955 North Carolina A&T University dedicated George Washington Carver Science Building on its campus.

1956 Simpson College dedicated Science Building in memory of George Washington Carver.

1965 Tuskegee Institute was designated by U.S. Congress as a National Historic Landmark.

GWC U.S. Polaris Submarine was launched at New Port News, Virginia.

1968 Alabama A&M University dedicated and named the George Washington Carver Complex.

1970 Graduate scholarships in memory of Carver established at Iowa State College.

1972 GWC Science Building was dedicated at University of Maryland, Eastern Shore, Princess Anne, MD.

1973 Carver was elected to Hall of Fame for Great Americans.

1974 U.S. Congress authorized establishment of Tuskegee Institute campus an a National Historic Site to include the campus an a Historic District, the Booker T. Washington Home, Grey Columns, and George Washington Carver Museum, all under the management of the National Park Service.

1977 Carver was enshrined in New York Hall of Fame for Great Americans.

1984 GWC Experiment Station dedicated at Tuskegee University.

University of Missouri College of Agriculture established GWC graduate fellowships.

GWC Room dedicated at University of Maryland Memorial Union.

1986   Carver was inducted into Iowa Inventors Hall of Fame.

1990   Carver was inducted into National Inventors Hall of Fame.

Carver was inducted into Alabama State Department of Archives and History.

Carver Memorial Research Farm was established by Lincoln University, Jefferson City, MO

Photos IV: Carver the Man

VIII.

CARVER DAY

George Washington Carver Day was celebrated for the first time on January 5, 1943. The following year, through the ardent efforts of Dr. Alma Illery, a bill which proclaimed this day of each year George Washington Carver Day, was passed by Congress and signed by the President of the United States. Dr. Illery saw that the life of this great man could further the brotherhood of man. George Washington Carver lives on as an inspiration to all.

To story of Dr. Alma Illery's life, built around the success of Camp Achievement and the national observance of "George Washington Carver Week," was recorded for worldwide broadcasts in 41 languages by William Meinhart and Joseph Petrglia of the United States Information Agenccy, Voice of America.

"Pittsburgh may well be proud of Dr. Alma Illery and of the hundreds of people who have helped her. The January 5th observance is a tribute to her as well as to the great George Washington Carver." (The late David L. Lawrence)

"We welcome to this noon's service of worship, Dr. Alma Illery, distinguished daughter and citizen of Pittsburgh, of whom it is true, as it was true of Dr. Carver, a life transfussed with passion for the enlarging and enriching of the lives of all - a liberator - a bridge from one race to another, of learning and understanding, and goodwill , and rejoicing in their commonwealth of opportunity and responsibility." (Emery Luccock, Chaplain, Heinz Chapel)

PROCLAMATION 2677

WHEREAS it is fitting that we honor the momery of George Washington Carver who contributed to the expansion of agricultural economy of the nation through his diligent research as an agricultural chemist; and

WHEREAS by a joint resolution approved December 28, 1945 (Public Law 290, 79th Congress), the Congress has designated January 5, 1946 as George Washington Carver Day and has authorized and requested me "to issue an proclamation calling upon officials of the Government to display the flag of the United States on all Government buildings on such day":

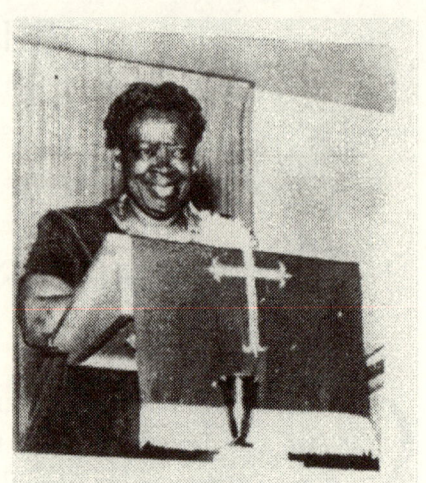

Dr. Alma Illery

NOW THEREFORE, I, HARRY S. TRUMAN, President of the United States of America, do hereby call upon officials of the Government to have the flag of the flag of the United States displayed on all Government buildings on January 5, 1946 in commemoration of the achievements of George Washington Carver.

IN WITNESS WHEREOF, I have hereunto set my hand and caused the seal of the United States of America to be affixed.

DONE at the City of Washington this 28th day of December, in the year of our Lord nineteen hundred and forty-five and of the Independence of the United States of America the one hundred and seventieth.

      HARRY S. TRUMAN [seal]

By the President:

  DEAN ACHESON
    Acting Secretary of State.

Photos V: Tributes and Honors

The Carver Museum

Carver and Franklin D. Roosevelt

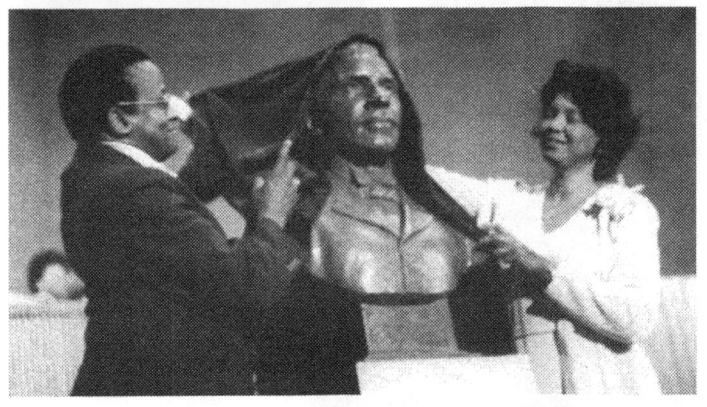

73

Below: Carver with Henry Ford

Right: U.S. Post Office honors Carver

Left: President Patterson delivers certificate of the Carver Research Foundation to Dr. Carver.

IX.

APPENDICES

A. Carver's Last Will and Testament

Filed January 21, 1943

THE STATE OF ALABAMA, MACON COUNTY

KNOW ALL MEN BY THESE PRESENTS, that 1, GEORGE WASHINGTON CARVER, being of sound mind and disposing memory, do hereby make and execute this my Last Will and Testament, expressly revoking any and all prior instruments, or codicils thereto, heretofore executed by me and purporting to be a Last Will and Testament:

It in my wish that all just debts due and owing by me, or all legal claims against my Estate, shall be paid in full. I, therefore, direct my Executor, as hereinafter named, to pay such legal debts as soon as practicable after the Probate of this Will. This to include expenses of last sickness, and funeral expenses, and also to include a simple and appropriate marker for my grave.

TWO: There are certain bequests that I desire to make to friends and acquaintances and I have listed the same on a separate paper attached hereto and signed by me. I direct my Executor to pay these bequests as specified, and in the amounts named, and to the persons authorized to receive same.

THREE: It is my wish that all property of every nature and description that I may own, or in which I may have an interest of any kind, at the time of my death, shall be turned over to my Executor as hereinafter named and by him controlled, handled, and disposed of, in the carrying out of the directions herein contained, and in the payment of the expenses connected with said Estate and the bequests specified in Paragraph Two. My purpose being, that all property of my Estate not needed in the payment of said expenses and legacies shall be finally turned over to a Trust Fund or Foundation and therein and thereby used and managed as more fully set out in Paragraph Five of this Will.

FOUR: There is now in process of organization a Corporation, not of a business character, as provided for by the Statues of the State of Alabama, the name of which will be "The George Washington Carver Foundation. " The said Foundation is established for the purpose of combining Research Laboratories and a Historical Museum, thus encouraging and carrying on the work started by me at Tuskegee Institute, and providing for the proper housing space for preserving same. Said Foundation also will have a more expanded and extended sphere of operation, but at the time of the execution of this Will it has not yet been fully carried into effect by complete organization. Assuming that said Corporation will be fully organized by the time that this Will becomes effective, I herewith and hereby make specific provisions for the disposition of my wordly effects, and for the carrying out of my intentions with reference to said Foundation.

FIVE: I hereby give, bequeath, and devise all property of every nature and description owned by me at the time of my death, and any property of any kind whatsoever in which I may have an interest, legal or equitable, at the present time or in the future, and all other things of value accruing to me, or to my estate, from any source whatsoever, to The George Washington Carver Foundation, as then or thereafter organized. out of said property, my Executor as hereinafter named is to pay the legal debts against my Estate as provided for in Paragraph one, and the Legacies as specified in Paragraph Two, and the residue of every nature and description is to be turned over to said Foundation to be held in Trust as follows:

FIRST: I have many more or less, valuable paintings or pictures, about a hundred in number, all representing my own handiwork. It is my hope and desire that these paintings will be preserved, and that the proper space be provided for the housing and preservation of same.

SECOND: My work at Tuskegee Institute has been largely confined to Chemical Research, and especially as applied to Agriculture. The modern definition of this line of work is Chemurgy, in which field of labor I have been to a great extent a Pioneer. It is my wish and desire that this work be encouraged and continued by the Tuskegee Institute, and for this purpose and to that end I direct that any such part of my property or the income therefrom, as in necessary or available, shall be so used as decided by those controlling said Foundation.

THIRD: The present Museum Building at Tuskegee Institute, which, an a tribute to my work, has been given my name, is now used for exhibiting the results and products of my labors, and it in my hope and desire that such will be Continued. I, therefore, direct that any portion of the income or the corpus of my Estate, at any time necessary or available, be used for the maintenance of said Museum Building so that it may be kept at all times in a fit and proper condition for the purposes intended.

FOURTH: It is my heartfelt hope and desire that The George Washington Carver Foundation, an referred to herein and as organized and operated, shall in the many years to come, contribute to some extent in the great task of building up a non-productive soil, especially throughout the South. I believe that it can be shown by actual demonstration to a people, now greatly discouraged and disheartened, that our soils can be built up and be made much more productive, and thus bring about a more ideal type of general prosperity and happiness for those whose labors are spent in the tilling of the soils. To this end and for that purpose I direct that such property as I may have and own, and as may be turned over to the Foundation as herein provided, be used and dedicated.

SIX: I hereby name, nominate and appoint F. D. Patterson, now President of Tuskegee Institute, or his successor in office at the time of the Probate of this Will, as my Executor, he to serve as such without bond of any kind whatsoever. I direct that this Will be turned over to such person for Probate, and that he immediately take charge of all of my effects, and take the necessary legal steps for the probate of this Will. In this connection I make known the fact that I have no relatives who have any claim upon my Estate. I, therefore, direct that, upon the filing of this Will for Probate, the Judge order publication be given in The Tuskegee News for a period of Three weeks, and that said notice may be directed to "Whomever it may concern", and that said notice give the time set for hearing the application to Probate the Will, and such other information as is usually required by law in such published notice, and that no other notice of said hearing be given.

I further direct that no inventory of my Estate be filed; that no Appraisers be appointed; that no final settlement be had by my Executor as herein named; and that the Probate Court require no other legal proceedings in connection with my Estate, as my Executor herein named can carry out my directions fully without any settlement in the Probate Court.

SEVEN: At the expiration of such period as the law fixes for the settlement of an Estate, and after the full period of time has passed from the date of published notice of the appointment of Executor, I direct that my Executor turn over to the Board of Trustees of The George Washington Carver Foundation all property of every nature and description that came into his hands as such Executor, and has not been expended in the payment of debts and legacies. At said time, the said Executor to make to said Board of Trustees a report and statement of the receipts and disbursements, and of the property on hand, its nature and form, the amount, and value, etc. upon the filing of said report, and the acceptance thereof by the Board of Trustees, my Executor shall be discharged from any further liability in connection with my Estate. My said Executor shall then file in the Probate office of Macon county, Alabama, a simple statement from the Board of Trustees, properly executed and authenticated, to the effect that said report and statement have been so filed and received, and that all property of my Estate has been turned over to said Board of Trustees. Upon the filing in said Probate Office of said Statement, my Estate is to be considered forever closed, without any further proceedings of any kind whatsoever in said Probate Court.

IN WITNESS WHEREOF, I, the said George Washington Carver, have hereunto set my hand and seal in the presence of two witnesses, who in my presence and at my request have hereunto attested this my Last Will and Testament, on this the Second day of February, 1940.

George Washington Carver   (seal)

ATTEST:

R. B. Stewart, Jr.
Joseph W. Williams

STATE OF ALABAMA
MACON COUNTY

I, Wm. Varner, Judge of Probate, in and for the county and state aforesaid, hereby certify that the above and foregoing instrument in writing was duly provided and admitted to Probate in this Court as the last Will and Testament of George Washington Carver deceased, and recorded, together with the Proof, in Book __ of Wills and Appraisements record   on the page thereof.   This the 22 day of Feby., 1943

Wm. Varner
Judge of Probate

Estate of George Washington Carver, Deceased

LETTERS TESTAMENTARY
STATE OF ALABAMA

MACON COUNTY
   PROBATE COURT

THE WILL of George Washington Carver, Deceased, of the said County, having been duly admitted to record, in said County, LETTERS TESTAMENTARY are hereby granted to F. D. Patterson, the Executor named in said will, who has complied with the requisitions of the law, and is authorized to take upon himself the execution of such Will.

Witness my hand and dated this 22nd day of February, 1943.

Wm. Varner
Judge of Probate

X.

CARVER BIBLIOGRAPHY

Books & Articles from Books

Adams, Russell L. *Great Negroes, Past and Present.*. 3rd. ed. Chicago: Afro-American Publishing Co., 1964, pp. 68-69.

Albus, Harry 1. *The Peanut Man,* Grand Rapids, Michigan: William B. Eerdmans Publishing Co., 1949.

Asimov, Isaac. *Breakthroughs in Science,* Boston: Houghton, 1959, pp. 172-176.

Bailey, Carolyn Sherwin. *Candle for Your Cake,* Philadelphia: Lippincott, 1952, pp. 3-12.

Bailey, Helen Miller. Forty *American biographies* New York: Harcourt, Brace and World, 1964, pp. 179-183.

Baker, Harry J. *Biographical* Sagas *of Will Power,* New York: Vantage, 1970, pp. 210-215.

Bolton, Sarah Knowles. *Lives of Poor Boys Who Became Famous,* New York: Thomas Y. Crowell, 1962, pp. 248-261.

Bontemps, Arna Wendell. *The Story of George Washington Carver* New York: Grosser and Dunlap, 1954.

Borth, Christy. *Pioneers of-Plenty: The Story of Chemurgy.*. Indianapolis: The Bobbs-Merrill Co., 1939, pp. 226-240.

Brandenberg, Aliki. A *Weed is a Flower: Ile Life of George Washington Carver,* Englewood Cliffs, New Jersey: Prentice-Hall, 1967.

Brimberg, Stanlee. *Black Stars.* New York: Dodd, Mead, 1974, pp. 37-50.

Bunker, John Gorley. *Liberty Ships.* Arno Press, New York Times Company, New York, 1980, pp. 41, 174.

Clark, Glenn. *Ile Man Who Talks with the Flowers: The Intimate Life Story of Dr. George Washington Carver.* St. Paul, Minn.: Macalester Park Publishing Co., 1939

Clifford, Harold Burton. *American Leaders,* New York: American Book Co., 1953, pp. 280-290.

Coates, Ruth Allison. *Great American Naturalists.* Minneapolis: Lerner Publications Co., 1974, pp. 89-95.

Cooper, A. C. and Palmer, C.A. *Twenty Modem Americans,* New York: Harcourt, Brace and Co., 1942, pp. 139-158.

Cox, Donald W., *Pioneers of Ecology.* Maplewood, New Jersey: Hammond, 1971, pp. 65-69.

Coy, Harold. *The Real Book About George Washington Carver,* **New** York: Garden City Books, 1951.

Curtin, Andrew. *Gall= of Great Americans,* New York: F. Watts, 1965, p. 13.

Dies, Edward Jerome. *Titans of the Soil,* Chapel Hill: University of North Carolina Press, 1949, pp. 180-181.

Dolin, Arnold. *Great Men of Science.* New York: Han Publishing Co., 1960, pp. 162-174.

Edwards, Ethel. *Carver of Tuskegee,* Cincinnati: Psych Press, 1971.

Elliott, Lawrence. *G. W. Carver: The Man Who Overcame,* Englewood Cliffs, New Jersey: Prentice-Hall, 1966.

Embree, E.R. *Thirteen Against the Odds.* New York: The Viking Press, 1944, pp. 97-116.

Epstein, Samuel and Beryl (Williams). *George Washington Carver. Negro Scientist,* Champaign, Illinois: Garrard Press, 1960.

Fisher, Dorothea Frances (Canfield). *And Long &member: Some Great Americans Who Have Helped Me.* New York: Whittlesey House, 1959, pp. 96-111.

Gilmartin, John G. and Skehan, Anna M. *Great Names in American*

History, Chicago: Laidlaw Brothers, 1946, pp. 343-350.

Goebel, Edmund Joseph, and others. *Builders of Our Country* Chicago: Laidlaw Brothers, 195 1, pp. 353-360.

Graham, Shirley and Lipscomb, Gap. *Dr. George Washington Carver. Scientist,* New York: Julian Messner, Inc., 1944.

Haber, Louis. *Black Pioneers of Science and Invention.* New York: Harcourt, Brace and World, 1970, pp. 73-85.

Hagedorn, Hermann. *Americans: A Book of Lives.* New York: The John Day Co., 1946, pp. 225-243.

Hall, Kenneth. *They Stand Tall: Life Stories of Fifteen Great Men and Women,* Anderson, Indiana: Warner Press, 1953, pp. 19-29.

Heath, Monroe. *Great Americans at a Glance.* Redwood City, California: Pacific Coast Publishers, 1955, p. 29.

Hill, Harvey Jay. *He Heard God's Whisper,* Minneapolis, Minnesota: Jorgenson Press, 1943.

Hines, Linda Elizabeth Ott. *Background to Fame: The Career of George Washington Carver.* 1896-1916. A dissertation, Auburn University, August 26, 1976, 221 pp.

Hoff, Carol. *They Served America,* Austin, Texas: Steck-Vaughn Co., 1966, pp. 33-37.

Holt, Rackham. *George Washington Carver: An American Biography* Revised edition. Garden City, New York: Doubleday, Doran and Co., 1963.

Hughes, Langston. *Famous American Negroes,* New York: Dodd, Mead, 1954, pp. 69-76.

Hunter, J. H. *Saint, Seer and Scientist,* Grand Rapids, Michigan: Zondervan Publishing House, 1939.

Johnson, Ruth 1. *Christians You Should Know,* Chicago: Moody Press, 1960, pp. 26-32.

Kelen, Emery. *Fifty Voices of the Twentieth Century,* New York: Lothrop, Lee, and Shepard, 1970, pp. 18-22.

Kitchens, John W. and Lynne B., eds. *Guide to the Microfilm Edition of the George Washington Carver Papers at Tuskegee Institute,* Tuskegee, Alabama: Tuskegee Institute, 1974.

Klein, Aaron E. *Hidden Contributors: Black Scientists and Inventors in*

America, Garden City, New York: Doubleday, 1971, pp. 130-143.

Law, Frederick Houk. *Great Lives.* New York: Globe Book Co., 1952, pp. 214-225.

Leipold, L. Edmond. *Famous American Negroes,* Minneapolis, Minnesota: T.S. Denison, 1967, pp. 49-54.

Lewis, Alethia. *A True Fairy Tale,* Boston: Christopher Publishing House, 1952.

McGuire, Edna. *They Made America Great,* New York: MacMillan, 1957, pp. 269-274

McCurry, Linda 0. *George Washington Carver: Scientist and Symbol.* New York: Oxford University Press, 1981.

McAleer, May Sponge and Ward, Lund K. *Armed With- Courage,* New York: Abingdon Press, 1957, pp. 39-54.

Member, David. *Wizard of Tuskegee: The Life of George Washington Carver,* New York: Crowell-Collier Press, 1967.

Martin, Fletcher, ad. *Our Great Americans: The Great Negro Contribution to American Progress.* Chicago: Gamma Corp., 1953, pp. 93-94.

Mason, Miriam Evangeline and Cartwright, William H. *Trail Blazers of American History,* Boston: Ginn, 1961, pp. 244-254.

Mayberry, BUD. *The Role of Tuskegee University in the Origin. Growth and Development of the Negro Extension System. 1881-1990* Tuskegee University Cooperative Extension System, October 1989, pp. 49-56.

Means, Florence Crannell. *Carver's George. A Biography of George Washington Carver,* Boston: Houghton, Mifflin Co., 1952.

Merritt, Raleigh H. *From Captivity to Fame: or, The Life of George Washington Carver* Boston: Meador Publishing Co., 1938.

Meyer, Edith Patterson. *Champions of the Four Freedoms,* Boston: Little, Brown, 1966, pp. 178-194.

Miller, Basil William. *George Washington Carver, God's Ebony Scientist,* Grand Rapids, Michigan: Zondervan Publishing House, 1943.

Miller, Basil William. *Ten Boys Who Became Famous,* Grand Rapids, Michigan: Zondervan Publishing House, 1946.

Moderow, Gertrude. *People to Remember* Chicago: Scott, Foresman, 1960, pp. 149-165.

Morris, Richard B., ed. *Encyclopedia of American History,* New York: Harper, 1961, p. 684.

Nelson, Rose Karen. *Carver of Tuskegee,* New York: New York Service Bureau for Intercultural Education, 1939.

Ovington, Mary White. *The Walls Came Tumbling Down,* New York: Harcourt, Brace and World, 1947.

Perry, John. *Unshakable Faith: Booker T. Washington & George Washington Carver* Multnomah Publishers, Sisters, Oregon.z

Pereira, Arty. They *Won Fame and Fortune.* Shahadera, Delhi, India: Hind pocket Books, 1943, pp. 104-109.

Phelps, George Allison. *Holidays and Philosophical Biographies* Los Angeles: House-Warven, 1951, pp. 25-31.

Powell, Lucille Rader. *Ten All-American Boys,* Philadelphia: Dorrance, 1971, pp. 41-55.

Pullen, Alice Muriel. *Despite the Colour Bar, the Story of George Carver. Scientist,* Student Christian Movement Press, 1946.

Richardson, B.A. *Great American Negroes.* New York, The Thomas Y. Crowell Co., 1956.

Rogers, Joel Augustus. *World's Great Men of Color.* New York: J. A. Rogers, 1947, vol 2, pp. 633-641.

Rothwell, Melvin Thomas. *George Washington Carver, a Great Scientist,* Chicago: Van Kampen Press, 1944.

Sawyer, L.A. and W.H. Mitchell. *The Liberty Ships.* The History of "Emergency" Type Cargo Ships constructed in the U.S. during the Second World War. Lloyds of London Press, Second Edition. New York and Great Britain, 1985, p. 135.

Schnittkind, Henry Thomas, and Schnittkind, Dana Arnold (pseud. Henry Thomas and Dana Lee Thomas). *Fifty Great Americans,* Garden City, New York: doubleday, 1948, pp. 348-357.

Schnittkind, Henry Thomas, and Schnittkind, Dana Arnold (pseud. Henry Thomas and Dana Lee Thomas). *Living Biographies of Great Scientists.* Garden City, New York: Perma Giants, 1950.

Simmons, Sanford. *Great Men of Science,* New York: Hart Book Co., 1955, pp. 51-57.

Smith, Alvin D. *George Washington Carver. Man of God.* New York: Exposition Press, 1954.

Southworth, Gertrude (Van Duyn) and Southworth, J. V. D. *Heroes of Our America,* Syracuse, New York: Iroquois Publishing Co., 1952, pp. 366-372.

Stevenson, Augusta. *George Carver.* Boy *Scientist,* New York: Ile Bobbs-Merrill Co., 1944.

Strong, Jay. *Famous Heroes of the* Ages. New York: Hart Publishing Co., 1958, pp. 182-191.

Thomas, Henry. *George Washington Carver.* New York: Putnam, 1958.

Edgar A. A *Biographical History of Blacks in America Since 1528.* New York: McKay, 1971, p. 266.

Washington, Booker T. *My Larger Education,* Garden City, New York: Doubleday, Page and Co., 1911.

Welch, Helena. *When They Were Children,* Nashville, Tennessee: Southern Publishing Association, 1965, pp. 103-106.

White, Anne Terry. *George Washington Carver: Ile Story of a Great American.* New York: Random House, 1953.

Yost, Edna. *Modern Americans in Science and Technology,* New York: Dodd, Mead, 1962.

Periodical References & Other Literature

Alba, N.C. "Carver," *Negro History Bulletin,* vol. 22 (December, 1958), p. 59.

Asimov, *Isaac* . "Carver, World in a Peanut," *Senior Scholastic,* vol. 74 (April 17, 1959), p. 11.

Beaubien, P. L. and Mattes, M. J. "George Washington Carver National Monument," *Negro History Bulletin. vol. 18 (November, 1954). pp. 33-38,*

"Black Leonardo," *Time,* vol. 38 (November 24, 1941), pp. 81-82.

Block, Maxine, ed. *Current Biography.* New York: H.W. Wilson Co., 1940, pp. 148-150.

Brown, R. W. "Research Laboratory, the Carver Foundation," *Science,* vol. 115 (May 23, 1952), P. 562.

Bunche, Ralph J. "The World Significance of the Carver Story," an address delivered at the 3rd annual Christian liberal arts festival at Simpson College, Indianola, Iowa (October 6, 1956).

"Carver National Monument to be Dedicated," *Christian Century*, vol. 70 (October, 1953), p. 765.

Childers, James Sacon. "A Boy who Was Traded for a Horse," *Tie American Magazine*, Vol. 114 (October, 1932), pp. 24-25.

Childers, James Sacon. "A Boy who Was Traded for a Horse," *Reader's Digest*, Vol. 30, no. 178 (February, 1937), pp. 5-9.

Clark, M. G. "Champions of Freedom," *Instructor*, Vol. 75 (January 1966), p. 111 "Dr. George Washington Carver Stamp," *Hobbies*, Vol. 52 (December, 1947), p. 126.

Elliott, L. "Beyond Fame or Fortune," *Reader's Dim*, Vol. 89 (May, 1965), pp. 259-262.

Fenner, M. S. and Soule, J. C. "George Washington Carver: The Wizard of Tuskegee," *National Education Association Journal*, Vol. 35 (December, 1946, pp. 580-581.

Ford, E. C. "Visit with George Washington Carver," Negro *History Bulletin*, Vol. 17 (October, 1953), p. 5.

German, B. "I Knew George Washington Carver," *Middle School*, Vol. 67 (May, 1953), pp. 14-15.

"Goober Wizard: Negro Scientist Turns Peanuts Into Vital Crop," *Lite=Digest*, Vol. 123 (June 12, 1937), pp. 20-2 1.

High, S. "No Greener Pastures, - *Reader's Dig=* Vol. 41 (December, 1942), pp. 71-74.

Kingsley, Ruth Reynard. "Me Peanut Story," *Highlights for Children.-* Vol. 22, no. 7 (August-September, 1967), pp. 22-23.

"Let Us Pray," *Christian Century*, Vol. 57 (April 3, 1940), pp. 445-446.

Love, R. L. "George Washington Carver: A Boy Who Wished to Know Why," *Negro History Bulletin*, Vol. 30 (January, February, March, 1967), pp. 13-15, 15-18, 15-19.

Luce, Claire B. "Saintly Scientist," *Vital Speeches*, Vol. 13 (February 1,

1947), pp. 241-245.
Lynch, James P. "Milking the Peanut," *Popular Science Monthly*, Vol. 96 (April, 1920), p. 74
Martorella, P. H. "Negro's Role in American History," *Social Studies*, Vol. 60 (December, 1969), pp. 318-325.
Massie, S. P. "George Washington Carver Story," *Chemistry*. Vol. 43 (September, 1970), pp. 18-21.
Mayberry, B.D. "The Tuskegee Movable School: A Unique Contribution to National and International Agriculture and Rural Development." Agricultural History, Vol. 65, No. 2, 1990, Agricultural History Society, pp. 85-104.
Moehlman, Arthur B. "George Washington Carver, Master Teacher," *Nation's Schools*, vol. 31 and 32 (March, 1943), p. 13.
"Monument to a Scientist Educator," *School Life*, vol. 40 (October, 1957), p. 10.
Moss, W. W. "Peanut Man," *Time*, vol. 29 (June 14, 1937), p. 54.
Moss, W. W. "Peanuts, How Scientist's 145 Varieties Helped Lowly Goober to Rise," *Popular Science Monthly*, vol. 102 (May, 1923), p. 68.
National Cyclopedia of American Biography. New York: James T. White and Co., 1947, vol. 33, pp. 316-317.
"A Negro Chemist,* *The Literary Digest* , vol. 83 (December 13, 1924), p. 25.
"New Research to Honor a Negro," *Newsweek*, , vol. 10 (September 6, 1937), p. 20.
"New Scientific Foundation and It's Founder," *Commonweal*, vol. 31 (March 15, 1940), p. 441.
"Our Country is Richer Because of Their Lives," *Instructor* , vol. 59 (May, 1950), p. 23.
Paris, L. "Wizard of the Soil," *Senior Scholastic*. vol. 72 (May 7, 1958), p. 11 "Peanut Man," Time, vol. 29 (June 14, 1937), p. 54.
"Prayer that Prevails," *Christian Century*, vol. 57 (May 8, 1940), p. 603.
Remmers, W. E. "Education of a Scientist," *Negro History Bulletin*, vol. 20 (March, 1957), pp. 130-32.

Stewart, O. "Carver of Tuskegee." *Scribners Commentator,* vol. 10 (May, 1941), pp. 12-16.

"Story of George Washington Carver," *Instructor.* vol. 62 (February, 1953), p. 15.

Taylor, A. W. "South Honors Negro Chemist," *Christian Century,* vol. 54 (May 26, 1937), p. 686.

"Temple of the Peanut: Negro Scientist Pledges $33,000 Project," News vol. 15 (February 26, 1940), pp. 42-44

Thomas, H. and Thomas, D. L. "Science Milestone: Excerpt from Fifty Great Modem Lives," *Science Digest,* , vol. 42 (September, 1957), pp. 85-90.

Tobias, C. H. "Some Outstanding Negro Christians," *Missionary Review of the World.* vol. 59 (June, 1936), pp. 297-298.

Tokunage, M. "Contribution of the Negro to Science and Invention," *Negro History Bulletin,* vol. 18 (April, 1955), p. 154.

Turner, G. C. *For Whom is Your School Named?" Negro *History Bulletin,* vol. 19 (February, 1956), p. I 11.

U. S. Congress. Senate. Committee on Public Lands and Surveys. *George Washington Carver National Monument, Missouri,* Joint hearings, 78th Congress, 1st Session, on S. 37, S. 312 and H. R. 647. Washington: U. S. Government Printing Office, 1943.

Wright, C. W. "George Washington Carver, an American Scientist, - *Journal Chemical Education,* vol. 23 (June, 1946), pp. 268-270.

Wright, Clarence W. "Negro Pioneers in Chemistry," *School and Society,* vol. 65 (February 1, 1947), pp. 86-87.

DR. B. D. (BENNIE DOUGLAS) MAYBERRY, a native Alabamian, graduated from Tuskegee Institute in 1937. Although George Washington Carver had officially retired by then, he would lecture from time to time as a guest of other professors and Mayberry considered it a pleasure and a privilege to be Carver's student.

Following his graduation with a Bachelor of Science degree, Mayberry made a name for himself as an innovative researcher and dynamic teacher at universities throughout the southeast. He began his graduate studies at Michigan State University in 1947, earning a master's degree as well as a Ph.D. in horticulture. When Mayberry became an instructor in the Horticulture Department at Michigan State, he was the first African American hired by the College of Agriculture. However, when Mayberry decided to purchase a home for his family in East Lansing, no real estate agent would sell to an African American. He decided to return to Tuskegee as the head of horticulture.

Mayberry was active in teaching, research, and administration at Tuskegee for more than thirty years, from 1950 until his retirement in 1981. He diligently pursued state and federal funding for the land grant universities (including Tuskegee) that the Morrill Act had established in 1890 to serve African Americans. Among many other noteworthy contributions, Mayberry also led the effort to establish Tuskegee's preforestry program and secured the funding that established the Macon County Community Action Program and the Tuskegee Model Cities Project.

www.ingramcontent.com/pod-product-compliance
Lightning Source LLC
Chambersburg PA
CBHW021021090426
42738CB00007B/861